The *Incredible* Submersible *Alvin* Discovers a *Strange* Deep-Sea World

Brad Matsen

Enslow Publishers, Inc.

40 Industrial Road   PO Box 38
Box 398   Aldershot
Berkeley Heights, NJ 07922 Hants GU12 6BP
USA   UK

http://www.enslow.com

Library of Congress Cataloging-in-Publication Data

Matsen, Bradford.
    The incredible submersible Alvin discovers a strange deep-sea world / Brad Matsen.
        p. cm. — (Incredible deep-sea adventures)
    Summary: Describes the 1977 journey of the submersible ship Alvin, in which three
explorers saw, for the first time, the strange marine life that dwells in hydrothermal
vents and other mysteries of the deep sea.
    Includes bibliographical references and index.
    ISBN 0-7660-2189-0 (hardcover)
    1. Underwater exploration—Juvenile literature. 2. Hydrothermal vents—Juvenile
literature. 3. Deep-sea ecology—Juvenile literature. 4. Alvin (Submarine)—Juvenile
literature. [1. Underwater exploration. 2. Hydrothermal vents. 3. Hydrothermal vent
ecology. 4. Alvin (Submarine) 5. Submarines (Ships) 6. Ecology.]
    I. Title. II. Series: Matsen, Bradford. Incredible deep-sea adventures.
    GC65.M38 2003
    551.46'07—dc21
                                            2002013823

Printed in the United States of America

10 9 8 7 6 5 4 3 2 1

To Our Readers: We have done our best to make sure all Internet Addresses in this book
were active and appropriate when we went to press. However, the author and
the publisher have no control over and assume no liability for the material available
on those Internet sites or on other Web sites they may link to. Any comments or
suggestions can be sent by e-mail to comments@enslow.com or to the address on
the back cover.

Photo Credits: National Oceanic and Atmospheric Administration (NOAA), pp. 4,
5, 6, 8–9, 10, 13, 14, 15, 19, 21, 22, 25, 26, 31, 33, 34, 36, 37; Wildlife Conservation
Society, p. 12; © Woods Hole Oceanographic Institution, pp. 3, 17, 20, 30, 35, 38,
40; © Calvin J. Hamilton, pp. 24, 28.

Cover Photos: NOAA; © Woods Hole Oceanographic Institution (background).

Contents

The submersible *Alvin* begins an exciting dive into the deep ocean.

CHAPTER 1

Stand By *to* Submerge

"*lvin, Alvin.* This is *Lulu.* You are cleared to dive." With those words the deep-diving submersible *Alvin* began one of its most famous missions to the bottom of the sea. *Alvin*'s pilot replied to the controller on the ship *Lulu.* "Roger, *Lulu. Alvin* is cleared to dive. Stand by to submerge."[1]

Alvin can dive thousands of feet beneath the surface of the sea. It carries a pilot and two observers. As *Alvin* goes down, the light from the sun fades away. Soon the sea is totally black, but *Alvin*'s searchlights help the crew see outside the little submarine.[2]

On this dive, the pilot and observers were searching for hydrothermal vents. These vents are holes in the

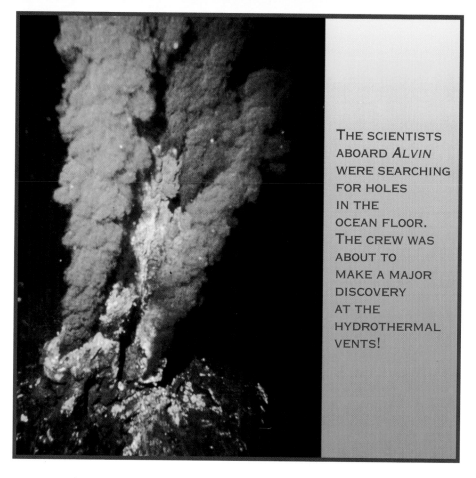

THE SCIENTISTS ABOARD *ALVIN* WERE SEARCHING FOR HOLES IN THE OCEAN FLOOR. THE CREW WAS ABOUT TO MAKE A MAJOR DISCOVERY AT THE HYDROTHERMAL VENTS!

bottom of the sea. Steam and warm water flow out of the holes. Scientists are very excited about exploring the areas around these strange undersea vents.

Alvin dropped deeper and deeper. Her crew heard the sounds of the electric propellers and fans. They heard the air inside the cabin flowing. The lights of the instrument dials and gauges glowed green. The little submersible had made many dives, but every dive is a new adventure. *Alvin* was diving again, on her way into the abyss.

Near the Bottom

In an hour, *Alvin* and her crew were one and a half miles (two and a half kilometers) below the ocean's surface (about 8,000 feet, or 2,400 meters). They were cruising near a ridge in the sea floor off the coast of South America. It was February 1977, and deep-ocean exploration was as exciting and challenging as space travel.

Almost every dive brought big surprises. On this one, the crew of *Alvin* discovered new forms of life! There in the searchlights were groups of strange animals living near the hydrothermal vents. This was almost as wild as if scientists had found life on Mars. *Alvin*'s excited pilot called up to the controller on *Lulu*.

"Isn't the deep ocean supposed to be like a desert?" the pilot asked.

"Yes," replied the controller up above.

"Well, there's all these animals down here."[3]

Life Around Hydrothermal Vents

Alvin's crew was shocked by what they saw. Crabs. Shrimp. Lobsters. Big pink fish. Huge clams and golden brown mussels. These creatures did not look exactly like their familiar cousins in shallow water, yet they were close enough to identify. But other creatures were nothing at all like anything anybody had ever seen. The strangest of these were wavy things that stood upright on the bottom in long tubes. Reddish tops poked out of the ends of the tubes. They look a little bit like giant lipsticks. There were

enormous thickets and groves of these creatures, called tube worms, all around the vents.[4]

The three people in *Alvin* that day were the first human beings to see most of those creatures. That voyage was planned for studying geology (the rocks and shapes of the bottom). There was no expert biologist on board *Alvin*. Still, the crew realized that they had discovered an entire new

NEAR THE HYDROTHERMAL VENTS, ANIMALS NEVER SEEN BEFORE WERE THRIVING. HUGE PATCHES OF TUBE WORMS WERE WAVING IN THE SEA.

world around the hydrothermal vents. Many more scientists would soon spend a lot of time in *Alvin* trying to figure out what was going on down there.

The water around the first vents they discovered was as warm as the water in a bathtub. The rest of the ocean around them was icy cold. *Alvin*'s crew captured some of the warm water and took it to the surface. When they opened the sample bottles on the deck of *Lulu*, the scientists smelled rotten eggs. That meant that the water contained a lot of sulfide, a chemical that comes from the inside of the earth through vents and volcanoes.[5]

The explorers who made the first dives to the hydrothermal vents off South America started a tradition of naming them. The ones where they saw the strange plants and animals for the first time are called Rose Garden, Garden of Eden, and East of Eden. Later names in the Atlantic Ocean were Lucky Strike, Broken Spur, and Snake Pit. Other favorites are Clam Acres, Snow Blower, and Genesis.[6]

The bottom of the sea would become more familiar and yet stranger with each of *Alvin*'s dives to the hydrothermal vents. Little *Alvin* proved herself to be the perfect craft for the exploration of the abyss. And wonderful stories unfolded.

CHAPTER 2

Diving *into* the Abyss

or most of human history, the bottom of the deep sea was a complete mystery. People waded in the water and stepped on sand and mud. They swam ten or twenty feet down to the bottom with their eyes open and saw fish, crabs, and other small creatures. Some ancient Greeks tried to walk deeper on the bottom by breathing through long straws.

About a hundred years ago, people began building diving helmets of leather and metal. The helmets had hoses leading to air pumps on the surface. Divers could only dive a hundred feet with them. Skin divers and scuba divers using masks, fins, and air tanks can only go down two or three hundred feet.[1]

Invention of the Bathysphere

Two explorers named William Beebe and Otis Barton made the first dives into the deep sea. They dove in a craft called a bathysphere. It was just a small hollow steel ball with two tiny windows. The bathysphere was lowered and raised on a cable from a ship. It could not actually land on the bottom of the sea, so the explorers could only study the middle of the water. In 1934, they set a record by descending a half mile; the record was not broken for fourteen years.[2]

Even though their craft was simple, Beebe and Barton were able to see the wonders of the deep ocean for the first time. They saw giant fish, strange jellylike animals, and creatures with flashing lights that had never before been seen by humans. They made notes and drawings of what they saw. Their expeditions into the depths changed the way scientists understand the ocean.[3]

Before scientists learned how to make deep dives, they thought the bottom of the sea was a wasteland with very little life. Finding weird fish and other strange creatures at great depths was a shock.

Invention of the Bathyscaphe

Beebe and Barton led the way a half mile down into the abyss in their bathysphere. Then submersibles called bathyscaphes were invented. (The word *bathyscaphe* means "deep boat.") Bathyscaphes do not need cables to go up and down. They really opened up the wonders of the deep.

William Beebe and Otis Barton stand on either side of their bathysphere. These explorers set a record when they dove a half mile into the deep sea.

By the late 1970s, explorers had been to the deepest place in the sea. Jacques Piccard and Don Walsh descended almost seven miles (eleven kilometers) into a trench named the Challenger Deep in the Pacific Ocean. Their bathyscaphe was named *Trieste*. When Piccard and Walsh got to the bottom of the Challenger Deep, they looked out their porthole. The bottom there looked like a kind of dark gray wasteland. To their surprise, however, they saw a small flatfish wriggle out of the mud and swim away.[4]

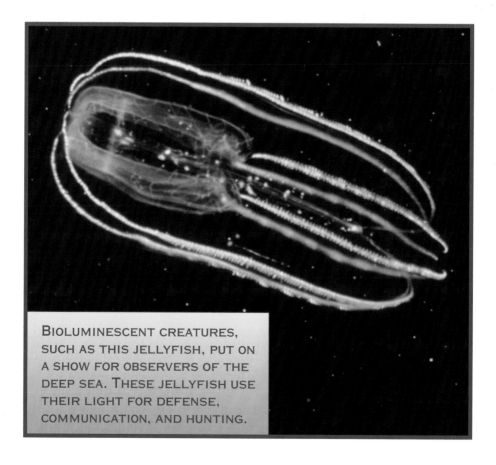

BIOLUMINESCENT CREATURES, SUCH AS THIS JELLYFISH, PUT ON A SHOW FOR OBSERVERS OF THE DEEP SEA. THESE JELLYFISH USE THEIR LIGHT FOR DEFENSE, COMMUNICATION, AND HUNTING.

Alvin's Design

Alvin is a bathyscaphe, too. It is a deep-water submarine with an engine. It is much stronger than the bathysphere, so it can go much deeper into the ocean. *Alvin* can also move forward and backward and land on the very bottom of the sea to a depth of 14,764 feet (4,500 meters). That's almost three miles down.[5]

Alvin's crew compartment is a sphere, just like the bathysphere's. However, it is made of titanium, a metal much stronger than steel. The walls of the sphere are almost two inches thick. *Alvin* has a metal and fiberglass hull around the sphere. It looks like a boat instead of a ball. It is 23 feet 4 inches long, 12 feet high, and 8 feet 6 inches wide. *Alvin* weighs 35,200 pounds, or about 17 metric tons, about as much as a school bus. Its maximum speed is 2 knots (about 3 miles an hour) and its cruising speed is ½ knot (about half a mile an hour). You can easily walk faster than *Alvin* can move through the water.[6]

Alvin is powered by five propulsion units called thrusters. The thrusters are electric and run on batteries. The lights and instruments also are powered by batteries. *Alvin* has two arms mounted on its front. They can be controlled from inside the cabin. Each arm can reach out about six feet and pick up things that weigh up to 250 pounds (113 kilograms).[7]

Alvin has cost $50 million to build and rebuild since it was launched in 1964. Every diving day costs about $25,000. *Alvin*'s home is the Woods Hole Oceanographic Institution in Massachusetts. It is officially owned by the United States Navy. The Office of Naval Research, the National Science Foundation, and the National Oceanographic and Atmospheric Administration all help pay the costs of *Alvin*'s missions.[8]

There is a funny story about how *Alvin* got its name. One of its designers was named Allyn Vine, so some people used parts of his first and last name when talking about the sub (*Al Vin*). But there was also a Christmas song that was popular at the time sung by a group called "Alvin and the Chipmunks." Something about the strange shape of the little submarine reminded people of a chipmunk, so the name stuck.[9]

Inside *Alvin*

Alvin's missions into the depths are usually eight hours long. If something goes wrong it can stay down longer. *Alvin* has enough air and food on board to keep its three crew members alive for seventy-two hours. The food for lunch on *Alvin* is pretty basic. It is usually something like peanut butter and jelly sandwiches, candy bars, fruit, and coffee. (*Alvin*'s crews envy the scientists on the French submersible named *Nautile*. Their mission food is a four-course hot lunch with wine.)

Alvin's cabin is about the size of a Jacuzzi. There is just enough room for three people. It looks like the cockpit of a jet fighter, because the pilot and observers are surrounded by instruments. The pilot sits on a small cushioned box. Passengers have to sit on the padded floor of the sphere. The pilot steers *Alvin* with a stick. The stick controls the thrusters.

Each observer has a small porthole for viewing the water outside. "The view is worth every tortured moment of discomfort it takes to hunker down, scrunch up, and peer out," says pilot Cindy Lee Van Dover.[10] The pilot and scientists also keep track of what is going on outside *Alvin* with video monitors. The cabin is lit by soft red light. Red light lets the pilots see better when they look into the sea. *Alvin* has powerful searchlights outside to light up the water and the sea bottom.

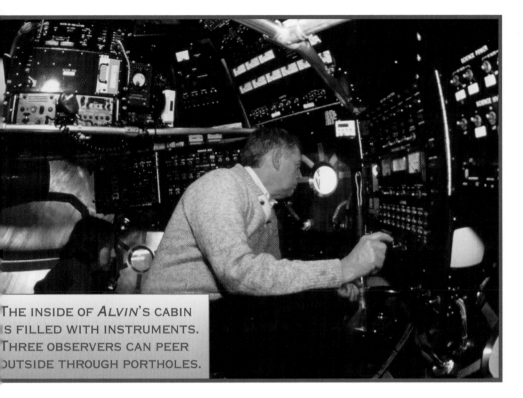

THE INSIDE OF *ALVIN*'S CABIN
IS FILLED WITH INSTRUMENTS.
THREE OBSERVERS CAN PEER
OUTSIDE THROUGH PORTHOLES.

Usually, geologists, biologists, and other scientists are the observers on *Alvin*'s dives. However, there are no special qualifications to be a passenger. You must only be healthy, unafraid of small spaces, and able to fit through the hatch. Everyone who rides in *Alvin* for the first time is shown how to bring it to the surface in the event of an emergency. First, you call an expert on the surface. Then you pull a special lever on the instrument panel. It releases heavy weights, called ballast. *Alvin* should rise to the surface.

If that does not work, there is a red handle under the floor cushions. That handle is the last hope if *Alvin* is trapped on the bottom. Nobody has ever had to use it. If it is used, though, the titanium sphere and its flotation material separates from the rest of *Alvin*'s ballast and the outside hull. The sphere rockets to the surface. No one knows if a person can survive such a trip.[11]

Alvin's Safety Record

"*Alvin* is a safe boat with an unsurpassed safety record," says Cindy Lee Van Dover, one of *Alvin*'s pilots and a well-known scientist. "Even so, no pilot can work the seafloor without wondering what it would be like to be trapped 4,000 meters down with only seventy-two hours to sort things out."[12]

Not one pilot or scientist has died in *Alvin* during a dive. Some, though, have had close calls. The greatest fear for an *Alvin* pilot is being trapped in a cave or being buried by an underwater mudslide. One time *Alvin* became wedged between two big rocks on the bottom. The pilot stayed calm but it took two hours for him to wiggle *Alvin* free. Another time, *Alvin* got stuck in thick mud on the bottom and barely escaped.

Diving in *Alvin* takes skill and courage. For those willing to make the trip, however, the wonders of the deep sea await them. After the discovery of the hydrothermal vents off South America, ocean exploration just kept getting better and better.

The Incredible *Alvin*

WEIGHT: 35,200 pounds, or 17 metric tons

LENGTH: 23 feet 4 inches (7.1 meters)

HEIGHT: 12 feet (3.6 meters)

WIDTH: 8 feet 6 inches (2.6 meters)

MAXIMUM DEPTH: 14,764 feet (4,500 meters)

MAXIMUM SPEED: 2 knots, or about 3 miles per hour

CRUISING SPEED: ½ knot, or about ½ mile per hour

POWER: Electric batteries

HOME PORT: Woods Hole Oceanographic
Institution, Massachusetts

LAUNCHED: 1964

Alvin's crew made another great discovery: chimneys on the ocean floor, spouting dark smoke. These formations are known as black smokers.

CHAPTER 3

The *Mysterious* Black Smokers

wo years after the first human contact with the creatures of the hydrothermal vents, *Alvin* and its crew discovered a stunning new realm of the abyss. On a dive two miles (three kilometers) deep off the coast of Mexico in 1979, the crew saw a giant chimney rising from the bottom of the sea. Inky black smoke was pouring out of the top.[1]

The smoking chimney was sixty-five feet (twenty meters) tall. That's about as high as a seven-story building. No one had ever seen anything like it before. *Alvin*'s pilot carefully brought the little ship right over to the scary-looking smokestack. As always, great discoveries call for great courage. The scientists were eager to know more about the mysterious object. They asked the pilot to use one of *Alvin*'s robotic arms to knock the top off the chimney and see what happened.

Remember, *Alvin* and her crew were two miles beneath the surface. They were far from help if something went wrong. Still, the pilot whacked at the smokestack with the robotic arm. Finally, the top came off. Inky black smoke shot out into the water even faster. Everything was still okay inside *Alvin*.

ALVIN WAS ABLE TO TAKE THE TEMPERATURE OF THE SMOKE THAT WAS POURING OUT OF THE BLACK SMOKER. THE TEMPERATURE AROUND THE BLACK SMOKERS IS OVER 650 DEGREES FAHRENHEIT (343 DEGREES CELSIUS).

Then the pilot used the robot arm to push a thermometer into the smoke that was coming out of the chimney. The crew could not believe their eyes. The readout said the temperature of the smoke was 91 degrees Fahrenheit (33 degrees Celsius). This could not be right, they thought. The water in the deep ocean is much, much colder than that. But they tried again and again and got the same temperature reading.

Later, on the surface, the crews of *Alvin* and *Lulu* examined the thermometer and its case. The mystery deepened. The thermometer's support structure was made of plastic that melts at 356 degrees Fahrenheit (180 degrees Celsius). The case had melted![2]

On later dives, *Alvin*'s crew got temperature readings of over 650 degrees Fahrenheit (343 degrees Celsius) around these strange formations. By then, they were calling them black smokers. And they knew better than to get too close to them in *Alvin*. The enormous heat could surely melt many parts of the little sub.

Just looking at a black smoker can be frightening. "Raw and powerful, black smokers look like the totem . . . [poles] of an inhospitable planet," said *Alvin* pilot Cindy Lee Van Dover. "I have often worked black smokers in *Alvin* and I never fail to be awed by them."[3]

Despite the danger of getting too close to a hot black smoker, *Alvin* and her crews have made hundreds of dives near them since they were first discovered in 1979. The weird chimneys and the creatures around them sent shockwaves through the world of deep-sea exploration.

Where Do Black Smokers Come From?

Scientists and explorers enjoy questions just as much as answers. The discovery of the hydrothermal vents, black smokers, and the animals that live nearby gave them plenty of questions. Where do they come from? What are they made of? What do they tell us about the earth? How can so much life exist so far down in the ocean? The list of questions is endless.

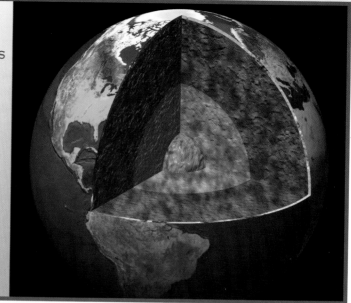

THE EARTH'S CRUST IS ROCK THAT SITS JUST BELOW THE SURFACE. IT IS ALSO BENEATH THE MUD ON THE BOTTOM OF THE OCEAN.

To understand where black smokers come from, it is helpful to imagine how the crust of the earth fits together. People who study this are called geologists. The crust is the rock that lies beneath the dirt and gravel on the land. The crust is also beneath the mud on the bottom of the ocean.

The whole earth is made up of giant plates of this crust that move around all the time. The crust moves very, very slowly. You cannot feel it moving except during an earthquake. The plates collide with each other and form mountains. In some places, the crust of one plate goes under another plate and returns to the inside of the earth.[4]

New crust is formed from molten rock inside the earth that spews out of cracks in the bottom of the ocean. These cracks are called spreading ridges. Hydrothermal vents appear along these ridges. Geologists were very excited when they found spreading ridges and hydrothermal vents. They were looking at the very beginning of the earth's new crust.

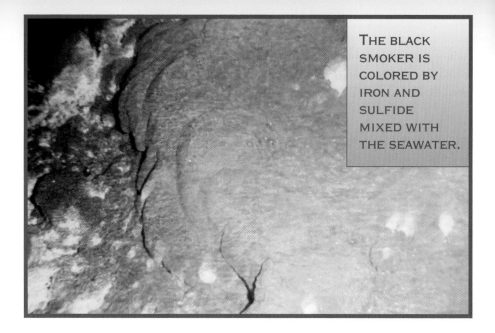

THE BLACK SMOKER IS COLORED BY IRON AND SULFIDE MIXED WITH THE SEAWATER.

Elements such as sulfur, gold, copper, iron, and zinc make the new crust. In some places, those elements mix with seawater. The water spews from the spreading ridge in plumes of super-heated water. Sometimes, minerals in the hot water form chimneys that rise from the bottom of the sea. The chimneys can spout black water, colored from iron and sulfide. These chimneys are called black smokers.[5]

There are many different kinds of black smokers. Some are tall chimneys. Some have mushroomlike tops. One black smoker in the Atlantic Ocean looks just like moose antlers. Its name is Moose. Another black smoker was named Beehive because that is what it looks like. One of the most famous black smokers is called Godzilla because it is so huge. Godzilla is on the ocean bottom off the coast of Washington. It is taller than a thirteen-story building and almost as wide as a football field.[6] Imagine cruising up to Godzilla in tiny little *Alvin*!

THERE IS AN AMAZING AND UNIQUE ECOSYSTEM AROUND HYDROTHERMAL VENTS. THE ANIMALS AND BACTERIA THAT LIVE THERE, SUCH AS THESE TUBE WORMS, ALL NEED SULFIDE TO LIVE. THIS IS THE ONLY PLACE IN THE WORLD WHERE THIS HAPPENS.

Life Around Black Smokers

The clams, mussels, and worms that live around black smokers and hydrothermal vents are very different from animals in other places. They thrive at high temperatures. They do not need sunlight to stay alive.

Food for animals around hydrothermal vents and black smokers comes from tiny organisms called bacteria. The animals that live near the vents, such as tube worms, clams, and mussels, would not exist if bacteria were not there to produce food for them. These bacteria are very different, too. They survive by absorbing chemicals, such as sulfides, instead of sunlight. Nowhere else in the world does this happen. All other life on the planet depends on energy from sunlight. But deep at the ocean's floor, life is powered by chemicals.

The animals and the bacteria that make their food depend upon each other for survival. This is called an ecosystem. The ecosystems around hydrothermal vents and black smokers are different from any other ecosystems on earth. Scientists were thrilled when they found totally new ecosystems.

Black smokers do not live forever. When they die, the bacteria and animals that live around them die, too. The bacteria also consume the bodies of dead clams, mussels, and worms. Then the spreading ridge often makes more black smokers. More animals and plants grow and thrive. The cycle continues.

CHAPTER 4

Becoming an *Alvin* Pilot

eologists like riding *Alvin* to black smokers to look at the formation of the earth's new crust. Biologists love their dives because they can visit animals unlike any others on the planet. Photographers and writers love diving in *Alvin* and returning to share the wonders of the deep with the rest of the world. All explorers of the abyss know the risks, but the rewards of discovery are too great to pass up.

But what kind of person wants to crouch inside a titanium ball day after day and command a dive to the bottom of the sea? What kind of person can remain calm when easing up to a black smoker 300 feet (91 meters)

high? What kind of person wants to become an *Alvin* pilot? A person like Cindy Lee Van Dover.

Before she became a submersible pilot, Van Dover studied to be a marine scientist. At first, she was interested in animals such as barnacles and oysters that live in mudflats. Then she went to work for a scientist who was studying crabs he found near hydrothermal vents. And so she fell in love with deep-ocean exploration.[1]

Beginning a Deep-Sea Career

Van Dover's first trip to the deep ocean was as an assistant aboard a research ship named *Melville* and a smaller craft named *Lulu*. *Lulu* carried *Alvin* to its dives. "Although I longed to dive in *Alvin*," Van Dover said, "I was content to be on one of the companion ships where the animals *Alvin* brought up from the vents were taken."

One day, though, she went over from *Melville* to *Lulu* to watch the men who took care of *Alvin*. "I envied the challenge of their job," she said. "I admired their experience, their confidence, their knowledge of how to work in the deep sea. . . . When the submersible surfaced that day, I rode the small boat and handled the lines during *Alvin*'s recovery. I dreamt then of just once being able to dive in *Alvin* as a scientist. I never dreamed that one day she would become my boat, my command."[2]

All of Van Dover's dreams and the dreams she did not dare to have came true. First, she earned her doctorate in deep-ocean ecology at the Massachusetts Institute of

CINDY LEE VAN DOVER STUDIED TO BE A MARINE SCIENTIST. THEN SHE BECAME ONE OF *ALVIN*'S TALENTED PILOTS.

Technology. One of the laboratories she worked at during her research was the Woods Hole Oceanographic Institution, *Alvin*'s home base.

The day after she received her doctorate, Van Dover started work for the team that was taking care of *Alvin*. "I knew it was the right thing to do, although it would take me once more off the beaten path to professional 'success' (as a scientist)," Van Dover said. "But what I have gained is worth far more."[3]

The Challenges of Becoming a Pilot

Many challenges lay ahead after Van Dover decided to work for the *Alvin* team. Her first job was writing a manual. It would help teach other people how to make their dives. She worked on the manual while *Alvin* was being rebuilt in the winter of 1989. When the little submarine was put back together again, Van Dover got a job with the sea-going crew as an assistant to the electrician. As a bonus, she was also given the title of pilot in training.[4]

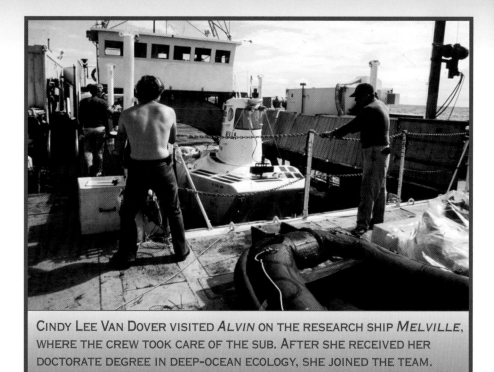

CINDY LEE VAN DOVER VISITED *ALVIN* ON THE RESEARCH SHIP *MELVILLE*, WHERE THE CREW TOOK CARE OF THE SUB. AFTER SHE RECEIVED HER DOCTORATE DEGREE IN DEEP-OCEAN ECOLOGY, SHE JOINED THE TEAM.

Van Dover learned to charge *Alvin*'s batteries and check and replace its many electrical cables. She also studied all of *Alvin*'s other systems so she would be able to deal calmly with an emergency during a dive, without any support from the ships on the surface. "This kind of challenge suited me," Van Dover said. "I was well motivated and certainly used to studying and figuring out how things work."[5]

Van Dover wanted to be an *Alvin* pilot more than anything else in the world, so she worked hard. "When I wasn't working on *Alvin*, I studied her so hard I thought my brain was going to burst," she says. "I would fall asleep with my pencil in hand, sketching out one more time from memory the power distribution system of the sub." Van Dover also dove in *Alvin* as a copilot and scientist.[6]

But then she hit a snag. Dudley Foster was the chief *Alvin* pilot and a former fighter pilot. One night, he took Van Dover aside and told her he was not sure she had the right stuff to become an *Alvin* pilot. She was devastated.

"*Alvin* pilots, I knew, must thrive under pressure," she said. "I listened to Dudley and vowed silently not to let this fresh cut in my self-confidence show. I demanded still more from myself. I was determined not to fail."[7]

Van Dover knew that one strike against her was that she had no special technical skills to offer as a pilot. She was a scientist. *Alvin* pilots had always been experts in mechanics, electronics, or other special jobs. She knew that a second strike against her was that she was a woman.[8]

The Golden Dolphins

Van Dover continued to work hard. Finally, in 1991, it was time for her final exams. To become a certified *Alvin* pilot, she had to take tests given by the U.S. Navy in San Diego. She had to prove she could dive and return to the surface. Underwater, she had to prove that she could operate all of *Alvin*'s controls. She had to show that she knew every system inside and out.

Van Dover's last test was an actual dive in *Alvin* with a check pilot. She passed with flying colors. The Navy awarded her its special pin given to research submersible pilots. The pin is two golden dolphin fish surrounding the famous sub *Trieste*, which took Piccard and Walsh to the bottom of the Challenger Deep in 1960.[9]

ALVIN HAS MANY SYSTEMS THAT THE PILOT MUST UNDERSTAND. PILOTS LEARN ABOUT ELECTRICITY AND MECHANICS.

Finally, Cindy Lee Van Dover was the first woman *Alvin* pilot.

Her orders: Go to Astoria, Oregon, and join *Atlantis*, *Alvin*, and their crews.

Her first dive as pilot in command: two miles down to a spreading ridge and an encounter with black smokers.

CHAPTER 5

Close *Encounter* with a Black Smoker

he beginning of a dive in *Alvin* is not as dramatic as the launch of a rocket to the moon. There is no countdown. No army of people supervises. There is no explosion or burst of flame. *Alvin* is not sleek. On the surface of the ocean, it is a clumsy little craft bobbing next to her mother ship.

In 1991, ten years after the discovery of black smokers, *Alvin* pilot Cindy Lee Van Dover quietly gave the command to submerge. She headed for the bottom of the ocean. On Van Dover's very first dive, she took two scientists to see the strange and frightening chimneys on the seafloor. She was on the way to the black smokers.

ALVIN HANGS FROM THE SHIP LULU. IT IS READY FOR ANOTHER ADVENTURE.

Each of you reading this book can go along on that dive with her. Just imagine that you are right there in *Alvin*. You can have a lot of confidence in your pilot. Van Dover has worked hard and received the best training in the world. You also can have a lot of confidence in the little submarine, *Alvin*. It, too, is a hard worker and a celebrity at the same time.

On the Way to the Bottom of the Sea

And now *Alvin* sinks into the sea. "As the descent begins, clear water quickly becomes aqua, then a deeper blue-green-black that has no name, then darker still until there is no color left at all. Only blackness," writes Van Dover about a dive in *Alvin*. "At 3,000 feet and deeper,

UNDERWATER, *ALVIN* ALLOWS ITS PASSENGERS TO VIEW A DARK WORLD FILLED WITH MYSTERIOUS SEA CREATURES, SUCH AS THIS SIPHONOPHORE.

splashes of bioluminescent light silently pass by like shooting stars."[1]

Inside *Alvin*, imagine you are sitting on the floor on black cushions. The light in the cabin is red, and you can see the green dials and video screens reflecting off the inside of the sphere. You are surrounded by glowing instruments and silver switches. Sensors and gauges monitor your depth, the temperature of the water outside, your downward speed, and the quality of the air you are breathing.

Pilot and observers share the tiny space with lunch boxes, thermos bottles of hot chocolate and coffee, cameras, a laptop computer, blankets, sleeping bags, flashlights,

WITH *ALVIN* BELOW THE SURFACE OF THE OCEAN, PASSENGERS ARE SURROUNDED BY GLOWING INSTRUMENTS AND THE DEEP, DARK SEA.

notebooks, and drawing pads. It is crowded, but you are too excited to notice as you pass the one-mile mark. You have entered the mysterious realm called the abyss.

You hear noise from the navigation system. It sounds like a tennis ball hitting a racket.[2] You also hear the hiss of oxygen flowing in the cabin. Every so often there is the "thunk" of a hydraulic valve opening or closing as Van Dover steers *Alvin*. She also sends signals to the mother ship to let them know everything is okay.

Down, down, down you go. For just a moment you become very frightened. You remember the safety meeting before the dive. You think about the red handle underneath

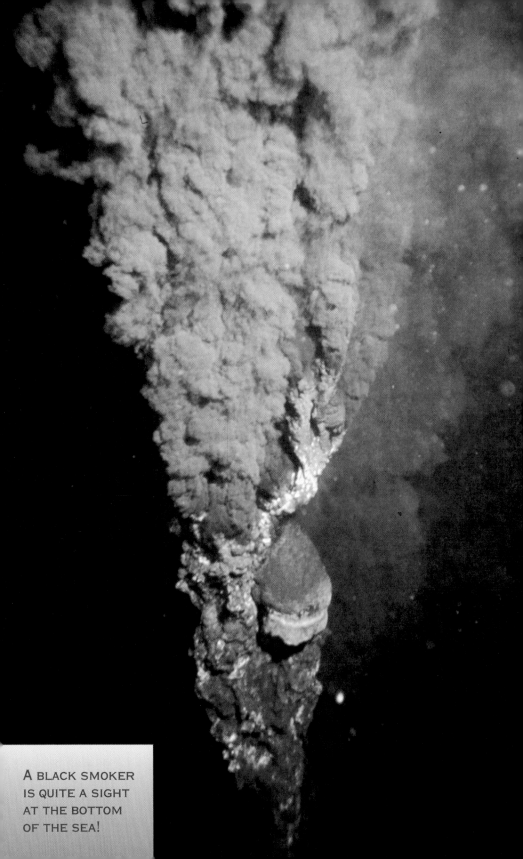

A BLACK SMOKER IS QUITE A SIGHT AT THE BOTTOM OF THE SEA!

the floor cushions. Your fear disappears when Van Dover switches on the searchlights outside *Alvin*. In their beams, you can see fish and jellyfish float by. They seem to slow down, and then the bottom is in sight. You hear thrusters hiss and klunk and feel *Alvin* begin to move forward instead of down.

Arriving at the Black Smoker

You have to scrunch down at the viewing ports to see outside. And you cannot believe your eyes. First you see only a tall dark shadow at the edge of *Alvin*'s pool of light. Then Van Dover brings you closer, and suddenly you see the base of something that looks like a giant tree. *Alvin*'s motors whirr. Valves thunk open and shut. You rise alongside the huge shape and see hunks of lava and rock stuck to what you now know must be a black smoker. Steam pours out of cracks in the chimney. A creature of some kind darts into and out of the light.

Your mission is to take a sample of the hot water from the top of the black smoker. This is tricky because the temperature can be hot enough to damage *Alvin*. Van Dover eases *Alvin* toward the top of the chimney. You think again of the red handle under the floor cushions.

Cindy Lee Van Dover carefully uses her controls to move *Alvin* closer and closer to the chimney. *Alvin* hovers near the open top of the black smoker. Van Dover uses the robotic arm to take some samples. Inside the cabin, the passengers seem to be holding their breath.

ONE OF THE TASKS DURING AN *ALVIN* DIVE IS TO TAKE A SAMPLE FROM A BLACK SMOKER.

You are two miles down on the bottom of the Pacific Ocean. Right outside the window—just inches from your face—is a real black smoker. You can feel *Alvin* bouncing in the rough water around the chimney. It takes only five minutes to finish the job, but it seems like hours. Finally, Van Dover and the scientists have recorded enough temperature samples.

Mission Accomplished

The dive is a success. And up you go, leaving the strange world of the black smokers below, in the depths of the Pacific Ocean. The dive was exciting, but you are almost glad to be on your way to the surface. As

Alvin rises, you eat your peanut butter and jelly sandwich and have some hot chocolate. Van Dover pays attention to piloting. The rest of you talk about the amazing black smoker you have just seen. This has been a once-in-a-lifetime thrill.

After her first dive, Van Dover went back to the bottom of the sea many times. She made forty-eight dives as *Alvin*'s pilot in command in five years. Many of those dives were to black smokers and hydrothermal vents. After she retired as a pilot, she continued her research in the deep ocean. She is now a professor at the College of William & Mary in Virginia.

"Although I loved diving the boat, I knew my real career was in research. If I am lucky, I get two or three dives a year in *Alvin*," she says. "But give me a dive when I am free to do whatever I want as if I were on a picnic on a lazy Sunday afternoon. I will spend that day in a field of black smokers, just looking."[3]

If you really want to dive to a black smoker, here is Van Dover's advice: "Work hard and love what you do."[4]

The next time you dive in *Alvin*, you might not have to use your imagination.

Chapter Notes

Chapter 1. Stand By to Submerge

1. Cindy Lee Van Dover, *The Octopus's Garden: Hydrothermal Vents and Other Mysteries of the Deep Sea* (New York: Addison-Wesley Publishing, 1996), p. 32.

2. Woods Hole Oceanographic Institution, *A History of Alvin*, n.d. <http://www.whoi.edu/marine/ndsf/vehicles/alvin/alvin_history.html> (September 18, 2002).

3. William Broad, *The Universe Below* (New York: Simon and Schuster, 1997), pp. 105–106.

4. Ibid., p. 106.

5. Ibid., p. 107.

6. Sean Chamberlin, *The Remarkable Ocean World*, "Black Smokers and Giant Worms," © 1999, <http://www.oceansonline.com/smokers.htm> (Feb. 23, 2002).

Chapter 2. Diving into the Abyss

1. Judith Gradwohl, ed., *Ocean Planet* (New York: Smithsonian and Harry N. Abrams, 1995), p. 99.

2. William Beebe, *Half Mile Down* (New York: Duell Sloan Pearce, 1951), <http://seawifs.gsfc.nasa.gov/OCEAN_PLANET/HTML/ocean_planet_book_beebe1.html> (Feb. 23, 2002).

3. Ibid.

4. William Broad, *The Universe Below* (New York: Simon and Schuster, 1997), pp. 54–55.

5. Woods Hole Oceanographic Institution, *A History of Alvin*, n.d. <http://www.whoi.edu/marine/ndsf/vehicles/alvin/spec_alvin.html> (September 18, 2002).

6. Ibid.

7. Broad, pp. 116–117.

8. Woods Hole Oceanographic Institution.

9. Ibid.

Chapter Notes

10. Cindy Lee Van Dover, *The Octopus's Garden: Hydrothermal Vents and Other Mysteries of the Deep Sea* (New York: Addison-Wesley Publishing, 1996), p. 35.

11. Ibid.

12. Ibid.

CHAPTER 3. THE MYSTERIOUS BLACK SMOKERS

1. William Broad, *The Universe Below* (New York: Simon and Schuster, 1997), p. 109.

2. Ibid., p. 110.

3. Cindy Lee Van Dover, *The Octopus's Garden: Hydrothermal Vents and Other Mysteries of the Deep Sea* (New York: Addison-Wesley Publishing, 1996), p. 110.

4. The American Museum of Natural History, *Black Smokers*, © 1997, <http://www.amnh.org/nationalcenter/expeditions/blacksmokers/black_smokers.html> (September 18, 2002).

5. Ibid.

6. Sean Chamberlin, *The Remarkable Ocean World*, "Black Smokers and Giant Worms," n.d. <http://www.oceansonline.com/smokers.html> (September 18, 2002).

CHAPTER 4. BECOMING AN *ALVIN* PILOT

1. State University of New York, Stony Brook, "Cindy Lee Van Dover," Biographical Sketch, January 26, 1998, <http://life.bio.sunysb.edu/marinebio/vandoverbio.html> (July 11, 2002).

2. Cindy Lee Van Dover, *The Octopus's Garden: Hydrothermal Vents and Other Mysteries of the Deep Sea* (New York: Addison-Wesley Publishing, 1996), pp. 15–18.

3. Ocean Adventures, *Meet the Scientists*, Cindy Lee Van Dover, n.d. <http://library.thinkquest.org/18828/data/sc_8.html?tqskip1=1&tqtime=0522> (July 11, 2002).

4. Ibid.

5. Van Dover, p. 26.

6. Ibid., p. 27.

7. Ibid., p. 26.

8. Ibid.

9. U. S. Navy, Office of Naval Research, Press Release, August 23, 1958, <http://www.onr.navy.mil/focus/ocean/vessels/submersibles6.htm> (July 11, 2002).

CHAPTER 5. CLOSE ENCOUNTER WITH A BLACK SMOKER

1. Cindy Lee Van Dover, *The Octopus's Garden: Hydrothermal Vents and Other Mysteries of the Deep Sea* (New York: Addison-Wesley Publishing, 1996), p. 32.

2. Ibid., p. 34.

3. Ibid., p. 101.

4. Ocean Adventures, *Meet the Scientists*, Cindy Lee Van Dover, n.d. <http://library.thinkquest.org/18828/data/sc_8.html?tqskip1=1&tqtime=0522> (July 11, 2002).

Glossary

abyss—The very deep parts of the ocean.

ballast—Disposable weights carried outside a bathyscaphe that can be released to make the craft lighter, so it can rise to the surface.

bathyscaphe—A manned research submarine that can operate without a cable to the surface, because its pilots can control its buoyancy.

bathysphere—A submersible chamber that is lowered and raised by a cable on the surface of the water.

biologist—A scientist who studies life-forms.

bioluminescence—Light produced by a living thing, such as a fish or jellyfish.

black smoker—A chimney rising from the floor of the sea through which hot water and steam pour into the sea.

buoyancy—The ability of a submarine or a diver to go up or down in the water.

carbon dioxide—The poison gas that is exhaled after a human or other animal breathes in air and removes most of the oxygen with its lungs.

diving suit—A suit made of rubber with a heavy metal helmet that allows a person to dive no deeper than sixty feet using air pumped from the surface through a hose.

Glossary

doctorate—The top university degree in science, called a Doctor of Philosophy, or Ph.D.

ecology—The study of living systems of plants, animals, and their surroundings.

geologist—A scientist who studies rocks and the formation of the earth's crust.

hydraulic—Powered by fluid pushed through hoses by a pump.

hydrothermal vent—A hole in the seafloor through which superheated water, steam, and lava flow into the ocean.

oceanographer—A scientist who studies the ocean, and its fish and other creatures.

scuba—*S*elf-*c*ontained *u*nderwater *b*reathing *a*pparatus. Scuba equipment lets divers swim underwater to depths of about two hundred feet using tanks of air carried on their backs.

spectrum—The colors visible to the human eye.

submersible—A craft for diving beneath the surface of the sea.

thruster—A valve that releases a burst of air or water to turn a bathyscaphe in the opposite direction.

titanium—A very strong, light metal used in building bathyscaphes, airplanes, and other modern machinery.

Further Reading

BOOKS

Demuth, Patricia. *Way Down Deep—Strange Ocean Creatures*. New York: Grosset & Dunlap, 1995.

Dipper, Frances. *Mysteries of the Ocean Deep*. Brookfield, Conn.: Copper Beech Books, 1996.

Earle, Sylvia. *Dive! My Adventures Under Sea*. Washington, D.C.: National Geographic Society, 1998.

Gibbons, Gail. *Exploring the Deep, Dark Sea*. Boston: Little Brown, 1999.

Kent, Peter. *Hidden Under the Sea*. New York: Dutton's Children's Books, 2001.

Markle, Sandra. *Pioneering Ocean Depths*. New York: Atheneum Books for Young Readers, 1995.

INTERNET ADDRESSES

American Museum of Natural History. *Black Smokers Expedition*. © 1997. <http://www.amnh.org/national center/expeditions/blacksmokers/>

University of Delaware College of Marine Studies and Sea Grant College Program. *Extreme 2000, Voyage to the Deep*. © 2000. <http://www.ocean.udel.edu/deepsea/home/home.html>

Woods Hole Oceanographic Institution. *History of Alvin*. <http://www.whoi.edu/marops/vehicles/alvin/alvin_history.html>

Index